A praying mantis lays her eggs

Terrariums

Jill Kalz

A⁺

Smart Apple Media

COPYRIGHT

❧Published by Smart Apple Media

1980 Lookout Drive, North Mankato, MN 56003

Designed by Rita Marshall

Copyright © 2002 Smart Apple Media. International copyright reserved in
all countries. No part of this book may be reproduced in any form without
written permission from the publisher.

Printed in the United States of America

❧Photographs by Jan A. Allinder, Kathy Adams Clark, Sally McCrae Kuyper,
Tom Myers

❧Library of Congress Cataloging-in-Publication Data

Kalz, Jill. Terrariums / by Jill Kalz. p. cm. — (Enclosed environments series)
Includes bibliographical references (p.).

❧ISBN 1-58340-106-7

1. Terrariums—Juvenile literature. [1. Terrariums. 2. Vivariums.] I. Title.

QH86 .K29 2001   578.07'3—dc21   00-051581

❧First Edition 9 8 7 6 5 4 3 2 1

Terrariums

CONTENTS

What is a Terrarium?

In 1829, Dr. Nathaniel Ward, a London doctor, held a glass jar up to his eye. The jar contained some soil and a moth cocoon. Dr. Ward was waiting for the moth to hatch. But he saw something strange. Tiny plants had poked out of the soil, even though the jar had been shut tight. How could they have grown in a sealed jar? Dr. Ward had created what we now call a **terrarium**. Terrariums are clear glass or plastic containers that hold soil and plants. Some terrariums also include

Ferns and mosses are typical terrarium plants

cold-blooded animals, such as lizards or turtles. Terrariums can be as small as a one-half gallon (1.9 l) fish bowl or as large as a 20-gallon (76 l) aquarium. Almost any clear container can be a terrarium. The top can be open or closed, depending upon what is growing inside.

Closed Terrariums

Terrariums are controlled **environments**. This means that the owner controls everything: air temperature, lighting, water, and food. Everything must be just right or the plants and animals will die. ❧ Closed terrariums are the most com-

mon terrariums. They are for plants only and usually look like

miniature gardens or forests. The plants are small and compact.

Some people use rocks, sticks, or tiny figures to create scenes.

Turtles lounge on a rock

Frogs are a terrarium favorite

They may build hills, cliffs, or even "rivers" of blue pebbles.

Taking Care of a Terrarium

Well-built closed terrariums need very little care.

Plants are protected from harmful fumes, insects, and temperature changes. Because the air is very **humid**, not all plants can live in this environment. Ferns, mosses, and small blooming plants work best. They love moist air. All plants chosen to put into a terrarium must need the same amount of moisture and light. A cactus, for example, would die in a humid environment. Cacti need dry, open terrariums with lots of sun. ❧ A

closed terrarium recycles its water. Plants draw water up

through the soil with their roots. Then they release it through

their leaves as **vapor**. This is called transpiration. Next, the

Moisture often collects on the glass of the terrarium

water vapor condenses and runs down the sides of the terrarium. It soaks back into the soil and the cycle begins again.

Some closed terrariums may not need watering for four to six months! ❧ Water is important, but so is good soil. A terrarium usually has four layers of growing material. The bottom layer is drainage material—pea gravel or crushed lava rocks—which keeps plant roots from rotting. ❧ Bits of charcoal make up the second layer. The charcoal absorbs smells caused by decaying plants. ❧ The third layer is peat

Terrariums were first used in London to protect plants from pollutants in the air.

moss. This layer holds the soil and keeps it from sifting to the

bottom of the container. ❧ The top layer is soil. Sterilized

soil works best because this soil has been heated to a high

Artificial lighting can be used in terrariums

temperature to kill bacteria. Regular garden soil can be too

heavy and muddy. It may also be unclean. Insects and molds

living in it can quickly make a terrarium sick. ❧ Closed ter-

rariums usually take care of themselves. But **Common**

some terrariums need a little help. Partially **closed terrar-**

ium plants

include ivy,

closed terrariums have covers that are easily **ferns, violets,**

begonias, and

removed. This allows owners to trim plants **small palms.**

that grow too fast or to feed animals. A few snails or lizards

may live in this kind of terrarium.

Snails may live in partially closed terrariums

What is a Vivarium?

Terrariums built especially for animals are called **vivariums**. Vivariums contain more than food and water bowls. They are fairly large so plants and animals have room to grow. Vivariums look just like an animal's natural environment. There are three basic types of vivariums: tropical, woodland, and desert. Each has its own animals, plants, lighting, food, and water. Some woodland vivariums may have water fountains or streams. Warming lamps may keep a desert vivarium

A skink hides behind leaves in a vivarium

hot and dry. Such conditions make vivariums good homes

for cold-blooded animals such as lizards, frogs, snakes, or sala-

manders. ❧ Like healthy terrariums, vivariums keep them-

selves clean. This task is made easier

by including a few hermit crabs. These

small creatures act like janitors, clean-

"Desertariums" are uncovered containers of dry, sandy soil, cacti, and other desert plants.

ing up plant decay and animal waste. In addition, microscopic

organisms in the soil help keep vivariums and terrariums

clean. ❧ Some people create terrariums as works of art.

Other people raise pets in them, while some simply want to

observe how nature works. Whatever the reason, terrariums let

people enjoy a little piece of the outdoors inside their homes

or schools.

Tarantulas make good terrarium pets

Recycling Water in a Terrarium

What You Need

A saucepan of water   Two oven mitts

A stove   An adult

A dinner plate

What You Do

1. Have an adult help you bring a saucepan of water to a boil. Watch how the water turns from a liquid to a gas (water vapor).
2. Put on your oven mitts. Hold the plate upside down just above the saucepan. Be careful! The steam is hot!
3. After about a minute, turn the plate over.

What You See

When you turn the plate over, it should be very wet. When the steam hit the plate, it condensed and turned back into water. If you hold the plate over the steam longer, water will soon drip back into the saucepan. This is what happens inside a closed terrarium. Plants take water from the soil. They release it as water vapor. This vapor then condenses and returns to the soil. All of the water is recycled.

A firebellied toad looks through the glass

I N F O R M A T I O N

Index

Words to Know

environments (en-VEYE-ren-ments)—surroundings of a person or animal

humid (HEW-mid)—air containing a high amount of water vapor

terrarium (tur-RARE-ee-um)—a transparent container that holds soil, plants, and sometimes small, cold-blooded animals

transpiration (TRAN-spi-RAY-shun)—the process of releasing vapor through pores

vapor (VAY-per)—a mist or gas

vivariums (vi-VAIR-ee-ums)—terrariums for animal life

Read More

Jes, Harald. *The Terrarium: A Complete Pet Owner's Manual*. New York: Barron's Educational, Inc., 1998.

Simon, Seymour, and Betty Fraser. *Pets in a Jar: Collecting and Caring for Small Wild Animals*. New York: Penguin Putnam Books for Young Readers, 1978.

Wardell, Randy A., and Judy Wardell. *Patterns for Terrariums and Planters*. Fort Lauderdale, Florida: Wardell Publications Inc., 1998.

Internet Sites

IcanGarden.com Home Page
http://www.icangarden.com/Latest From/RealDirt/rd139.htm

Great Plant Escape: University of Illinois Extension

http://www.urbanext.uiuc.edu/gpe/index.html

Kid's Valley Webgarden
http://www.arnprior.com/kidsgarden/index.htm